FIELD GUIDES
FOR KIDS

TREES
Andrea Debbink

Abdo Reference

An Imprint of Abdo Publishing | abdobooks.com

CONTENTS

WHAT IS A TREE?

Trees cover nearly 30 percent of Earth's land surface. They can thrive in many different environments: tundra, deserts, rain forests, and mountains. Sometimes, people take trees from their native areas and grow them elsewhere. Trees can be used to make a wide variety of everyday products, such as paper, clothing, candles, shampoo, tires, toothpaste, and much more.

WHAT DO TREES DO?

Trees do a lot of work to keep Earth healthy. People sometimes call them the "lungs of the Earth." That's because they can do many things, including:

- Filter polluted air.

- Produce oxygen.

- Provide food and shelter for animals, insects, and people all over the world.

- Store carbon in their leaves and trunks. Because of a tree's ability to store carbon, there's growing evidence that planting more trees—and stopping the destruction of forests—can help counter the negative effects of climate change.

CATEGORIES OF TREES

Scientists often define a tree as a living organism that has a root system, woody stem, and branches that form a crown. True trees typically grow to be at least 12 feet (3.6 m) tall. They are also perennial plants, which means they continue to grow year after year.

Trees belong to two main categories of plants: gymnosperms and angiosperms.

4

- A gymnosperm is a plant that doesn't flower and reproduces by creating exposed seeds—these are called seed cones. Conifers, cycads, ginkgophytes, and gnetophytes are all gymnosperms.

- Many more trees belong to the angiosperm category of plants. An angiosperm is a flowering plant that produces fruit containing seeds. Apple trees, maples, and palms are all angiosperms.

- There is a third category of tree-like plants called pteridophytes. But these aren't true trees. Most of these plants are shrubs or ferns.

TREE IDENTIFICATION

When identifying a tree, it helps to pay attention to the following characteristics:

- Height: Most tree species have an average range.

- Overall shape: Trees can grow in many different shapes, from tall and narrow to large canopies with multiple trunks.

- Bark: The bark is the protective layer on the tree's trunk and branches. It can be a variety of textures and colors.

- Leaves: One of the easiest ways to identify a tree is to examine its leaves. Leaf shape, color, texture, and arrangement can give people clues about the tree's identity.

- Flowers, fruit, and seeds: If a person examines a tree during the spring or fall, he or she can also use the tree's flowers, fruit, or seeds to help identify it.

HOW TO USE THIS BOOK

Tab shows the tree category.

The tree's scientific name appears here.

ANGIOSPERMS

ALOE YUCCA *(YUCCA ALOIFOLIA)*

The aloe yucca is also known as the Spanish bayonet tree.
  it has long, sword-like leaves that can be
  leaves grow along the tree's entire trunk.
  and often fall off once they die. The
  f the tree is the very top, where a crown of
  ow in a sunburst shape. The aloe yucca's
leathery seeds begin as large, white flowers.

The tree's common name appears here.

FUN FACT

Angiosperms are the largest—and most diverse—group in the plant kingdom. This group includes
  000 species of flowering plants.

Fun Facts give interesting information related to the tree.

HOW TO SPOT

Height: 5 to 20 feet (1.5 to 6 m)
Leaves: Pointed leaves; 2 feet (0.6 m) long
Fruit: Fleshy, green to black pods; 3 to 4 inches (7.6 to 10 cm) long
Range: Southern United States
Habitat: Coastal areas

YUCCA PLANTS

All yuccas are evergreen plants that have thick, blade-like leaves and that usually grow in warm, dry climates. There
  species of yucca plants worldwide, but most

Sidebars provide additional information about the topic.

30

JOSHUA TREE *(YUCCA BREVIFOLIA)*

The Joshua tree is the tallest of all yucca plants. This slow-growing tree has long, thin leaves that grow i... spiky clusters at the ends of its branches. The tru... branches of the tree are usually covered with a br... shaggy layer of dead leaves. In the spring the Jos... produces pale-green flowers that transform into t... egg-shaped seeds.

This paragraph provides information about the tree.

HOW TO SPOT

Height: 20 to 40 feet (6 to 12 m)

Leaves: Spike-like; 12 inches (30 cm) long

Fruit: Egg shaped; 2 to 4 inches (5 to 10 cm) long

Range: Southern California

Habitat: Desert

How to Spot **features give information about the tree's height, leaves, fruit/seeds, range, and habitat.**

Images show the tree.

FUN FACT
The Joshua tree is a very rare tree. It grows only in one small region of the world—the Mojave Desert.

31

AMERICAN ARBORVITAE
(THUJA OCCIDENTALIS)

The American arborvitae is a popular evergreen tree. It's often used in landscaping and might be seen in local parks. It's slow growing and offers food and shelter to many kinds of wildlife, including birds, deer, rabbits, and red squirrels. When its tiny, scale-like leaves are crushed, they release a strong piney scent.

HOW TO SPOT

Height: 30 to 45 feet (9 to 14 m)
Leaves: Tiny, scale-like leaves
Seeds: Cones; 0.5 inch (1.3 cm) long
Range: Northeastern United States and Canada
Habitat: Cool, damp soil near water

WHAT ARE EVERGREENS?

Evergreens are trees that lose and regrow their leaves gradually over a long period of time, sometimes years. This means they often appear green even during seasons when other trees shed their leaves.

GIANT ARBORVITAE *(THUJA PLICATA)*

The giant arborvitae is also known as the western red cedar. It grows so tall that the base of its trunk often is buttressed. This means it grows extra-thick folds to help support its height. This tree's crown is often cone shaped and has branches that arch upward. Its tiny, scale-like leaves have white marks on their undersides, and its cones are tiny too. The cones also have a dozen hook-shaped scales.

HOW TO SPOT

Height: 60 to 130 feet (18 to 39 m)

Leaves: Tiny scales

Seeds: Cones; 0.5 inch (1.3 cm) long

Range: Pacific northwestern United States

Habitat: Most common in moist coniferous forests

FUN FACT
The name *arborvitae* is Latin for "tree of life."

BALD CYPRESS *(TAXODIUM DISTICHUM)*

The bald cypress is a large, swamp-loving tree. It thrives in places where other trees might struggle. Near the base of the bald cypress are pointed growths called cypress knees. These are cone-shaped roots that rise up out of the water. The bark of the bald cypress is reddish brown and peels in vertical strips. In the fall, the tree can also be recognized by its orange-brown leaves.

HOW TO SPOT

Height: 80 to 100 feet (24 to 30 m)

Leaves: Flat needles; 0.75 inch (1.9 cm) long

Seeds: Round cones; 1 inch (2.5 cm) long

Range: Southeastern United States

Habitat: Swamps and near streams

MEDITERRANEAN CYPRESS
(CUPRESSUS SEMPERVIRENS)

In the wild, a Mediterranean cypress grows in a broad, horizontal shape. But it's more widely recognized in its cultivated form: a narrow column. A Mediterranean cypress's leaves are very small and scale-like and have almost no scent. Its branches grow in irregular-shaped sprays.

FUN FACT
The Mediterranean cypress is also called the Italian cypress. It has been planted in Italian gardens since the Renaissance era. That's more than 400 years ago!

HOW TO SPOT

Height: 40 to 70 feet (12 to 21 m)
Leaves: Scales
Seeds: Egg-shaped cones; 1.5 inches (3.8 cm) long
Range: Mediterranean region and southwestern Asia
Habitat: Rocky soil; often found in the mountains

WILDFIRE AND THE MEDITERRANEAN CYPRESS

The Mediterranean cypress is more resistant to fire than other tree species. That's because it usually doesn't accumulate dead branches and leaf clutter. In July 2012, a wildfire raged outside of Andilla, Spain. Tens of thousands of acres of trees were burned. But almost 1,000 Mediterranean cypress trees appeared untouched. Scientists are still trying to figure out how these cypress trees survived the fire.

EASTERN RED CEDAR
(JUNIPERUS VIRGINIANA)

The eastern red cedar is a type of juniper tree. Like most junipers, it can take many forms. Most of the time it grows tall and upright like a tree. But at other times the eastern red cedar can look like a short, squat shrub. It can even look like a creeping plant that grows along the ground. No matter what shape the tree takes, its needles and seeds always look the same. Its tiny cones grow inside a round, fleshy fruit that looks like a tiny blueberry when it's ripe. These cones are called juniper berries.

HOW TO SPOT

Height: 40 to 50 feet (12 to 15 m)

Leaves: Pointed, scale-like leaves that are green or blue green

Seeds: Round, berry-like cones; 0.25 inch (0.63 cm) long

Range: Eastern United States and Canada

Habitat: Forests and rocky slopes

FUN FACT
The eastern red cedar is called the pencil cedar. That's because its wood is often used to make pencils.

12

DROOPING JUNIPER
(JUNIPERUS RECURVA)

The drooping juniper gets its common name from the way it looks. The tree's long branches bend slightly backward, and its needle-like leaves point to the ends of these curves. Underneath the thick branches, the tree's narrow trunk is covered with reddish-brown bark. This bark regularly peels off in strips. The round, fleshy cones of the drooping juniper are nearly black when ripe.

HOW TO SPOT

Height: 10 to 60 feet (3 to 18 m)

Leaves: Needle-like; 0.25 inch (0.63 cm) long

Seeds: Round, berry-like cones; 0.3 inch (0.76 cm) long

Range: Southwest China, especially in the Himalayan Mountains

Habitat: Mountains

JUNIPER BERRIES

There are about 60 species of junipers. All junipers produce berries, but only three species of North American junipers have edible berries. Most juniper berries will make people and animals sick. The edible berries can be used in cooking.

BALSAM FIR *(ABIES BALSAMEA)*

The balsam fir is an evergreen tree. It has a recognizable cone shape. Its vibrant green color makes it a popular Christmas tree. In the wild it can grow up to 125 feet (38 m) tall. Unlike spruces and pines, fir trees have short, blunt needles that are somewhat soft. Similar to other firs, the balsam fir's cones grow upright and only appear on the top branches of the tree.

HOW TO SPOT

Height: 40 to 60 feet (12 to 18 m)

Leaves: Green needles; 0.75 inch (1.9 cm) long

Seeds: Cones; 3 inches (7.6 cm) long

Range: Central to eastern United States and Canada

Habitat: Moist woodlands and swamps

FUN FACT

The balsam fir is named for the syrupy resin that drips from its bark. This strong-smelling resin is called balsam. It has been used for medicine, glue, and varnish.

WESTERN HEMLOCK
(TSUGA HETEROPHYLLA)

The western hemlock is the tallest of the hemlock trees. Although hemlocks tend to have slender shapes, they're still usually broader than most spruces and firs. All the tips of the western hemlock's branches bend slightly downward. Like most conifers, the tree's needles are green, but they have white stripes on their undersides.

HOW TO SPOT

Height: 75 to 120 feet (23 to 36 m)

Leaves: Flat needles; 0.75 inch (1.9 cm) long

Seeds: Egg-shaped cones with rounded scales; 0.75 inch (1.9 cm) long

Range: Western United States and Canada

Habitat: Forests

JAPANESE LARCH *(LARIX KAEMPFERI)*

The Japanese larch is a tall tree that has soft, blue-green needles. In the fall, these needles turn bright yellow before they drop to the ground. The tree's bark is reddish brown and appears rough and scaly.

FUN FACT
Larches are the only members of the pine family that lose their needles each winter.

HOW TO SPOT

Height: 60 to 80 feet (18 to 24 m)
Leaves: Soft needles; 1.5 inch (3.8 cm) long
Seeds: Egg-shaped cones; 1.25 inch (3 cm) long
Range: Central Japan
Habitat: Mountains

LIMBER PINE *(PINUS FLEXILIS)*

The limber pine is named for its soft, flexible branches. The rubber-like branches can be bent in nearly any direction without breaking. Similar to other pine trees, the limber pine's long needles grow in clusters of three to five. Young trees will have smooth, pale-gray bark while older trees will have gray bark that appears scaly or has thick ridges.

HOW TO SPOT

Height: 35 to 50 feet (11 to 15 m)

Leaves: Curved needles; 2.25 inches (5.7 cm) long

Seeds: Cones; 5 inches (13 cm) long

Range: Rocky Mountains from Canada to New Mexico

Habitat: High mountains

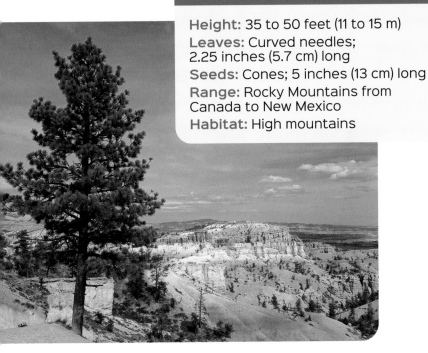

PINE TREE CONES

Every species of pine tree produces cones. However, they differ in size and shape. Most pine cones are egg shaped and have woody scales that grow in a spiral. But pine cones can also be round or long and thin. It takes two to three years for most pine cones to reach maturity and release their seeds. Every few years, pine trees produce extra-large crops of cones to increase the survival rates of the seeds.

SCOTS PINE *(PINUS SYLVESTRIS)*

Most Scots pines have curved trunks and asymmetrical shapes. The tree's bark changes color and texture from the ground to the crown. The bark on the lower part of the tree's trunk is dark gray with thick ridges. The bark on the upper part of the trunk is reddish orange and flaky. In the summer, the Scots pine's needles are blue gray but change to yellowish green in the winter.

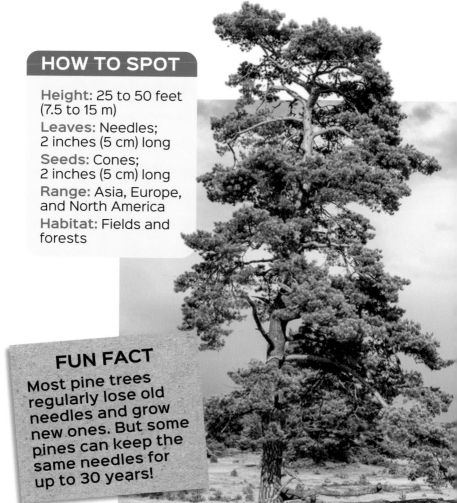

HOW TO SPOT

Height: 25 to 50 feet (7.5 to 15 m)

Leaves: Needles; 2 inches (5 cm) long

Seeds: Cones; 2 inches (5 cm) long

Range: Asia, Europe, and North America

Habitat: Fields and forests

FUN FACT

Most pine trees regularly lose old needles and grow new ones. But some pines can keep the same needles for up to 30 years!

NORWAY SPRUCE *(PICEA ABIES)*

Spruce trees are the only conifers that have needles that grow in a single shoot rather than in clusters. The Norway spruce's needles are sharp and stiff. Its large cones are pendulous, which means they hang downward and can swing back and forth. The tree's bark is usually a coppery brown and has paper-like scales.

HOW TO SPOT

Height: 50 to 80 feet (15 to 24 m)

Leaves: Needles; 0.75 inch (1.9 cm) long

Seeds: Cones; 6 inches (15 cm) long

Range: Europe

Habitat: Mountain forests

COAST REDWOOD
(SEQUOIA SEMPERVIRENS)

The coast redwood is the tallest living thing in the world. This mighty evergreen tree has relatively short, feathery branches that are angled upward. The tree's trunk is covered with thick, reddish-brown bark that has deep grooves and ridges. Despite the coast redwood's size, its cones are somewhat small. They're only about 1 inch (2.5 cm) long once they reach maturity.

HOW TO SPOT

Height: 150 to 367 feet (45 to 112 m)

Leaves: Flat needles; 1 inch (2.5 cm) long

Seeds: Cones; 1 inch (2.5 cm) long

Range: Pacific coast of the United States, such as California and Oregon

Habitat: Low slopes

FUN FACT

The tallest tree in the world is a coast redwood named Hyperion. It's 379.5 feet (115.7 m) tall and about 600 years old!

CUTTING DOWN FORESTS

About 200 years ago, coast redwoods covered around 2 million acres (810,000 ha) of land along the Pacific coast of the United States. When the California Gold Rush began in 1849, many of these ancient trees were cut down for lumber. In early 2020, only 5 percent of this original coast redwood forest was left.

GIANT SEQUOIA
(SEQUOIADENDRON GIGANTEUM)

The giant sequoia is a towering tree like its cousin, the coast redwood. The branches of the giant sequoia start very high on the tree's trunk. They're angled down and form rounded clumps. The giant sequoia's cones are twice the size of the coast redwood's and take twice as long to grow, maturing in about two years.

HOW TO SPOT

Height: 100 to 250 feet (30 to 76 m)

Leaves: Sharp, scale-like leaves; 0.3 inches (0.76 cm) long

Seeds: Barrel-shaped cones; 3 inches (7.6 cm) long

Range: California

Habitat: West-facing mountain slopes

JAPANESE UMBRELLA PINE
(SCIADOPITYS VERTICILLATA)

The small Japanese umbrella pine can often look like a shrub because of its many trunks. The tree's needles grow in clusters at the tips of its branches, fanning out into an umbrella shape. Like many other conifers, the umbrella pine's cones are soft and green at first but gradually harden and turn brown as they reach maturity.

HOW TO SPOT

Height: 25 to 40 feet (7.5 to 12 m)

Leaves: Grooved needles; 4.75 inches (12 cm) long

Seeds: Egg-shaped cones; 3 inches (7.6 cm) long

Range: Japan

Habitat: Mountains

COMMON YEW *(TAXUS BACCATA)*

In the wild, the common yew is a medium-sized tree with slightly drooping branches. Its trunk is covered in smooth, purple-brown bark. In farm fields or yards, the common yew is often trimmed into a hedge. The tree's flat, needle-like leaves grow in two rows along each twig.

FUN FACT
Nearly every part of the common yew is poisonous to humans and animals if eaten.

HOW TO SPOT

Height: 35 to 100 feet (11 to 30 m)

Leaves: Flat, needle-like leaves; 1.25 inches (3 cm) long

Seeds: Seed is enclosed in a red, fleshy fruit called an aril; 0.4 inches (1 cm) long

Range: Africa, southwestern Asia, and Europe

Habitat: Chalky or limestone soils

JAPANESE SAGO PALM
(CYCAS REVOLUTA)

The Japanese sago palm has a short, shaggy trunk that takes a very long time to grow. The trunk is topped with green leaflets that grow outward to form a feathery, green crown. Like all cycads, Japanese sago palms are dioecious. That means there are separate male and female plants.

HOW TO SPOT

Height: 3 to 20 feet (0.9 to 6 m)

Leaves: Glossy, green fronds; 3 to 7 feet (0.9 to 2 m) long

Seeds: Upright cones; 12 to 18 inches (30 to 45 cm) long

Range: Southern Japan

Habitat: Tropical forests

FUN FACT

Cycads have been around since the time of the dinosaurs. They've been growing on the planet for around 280 million years.

SILVER CYCAD *(CYCAS CALCICOLA)*

Young, flat fronds on the silver cycad are silver in color. As these plants get older, their fronds can become a deep green above, but they always have a white color below. On the plant's fronds are between 210 and 410 narrow leaflets. In 1978, Australian botanist John Maconochie discovered the species.

HOW TO SPOT

Height: 10 feet (3 m)
Leaves: 24 to 50 inches (60 to 130 cm) long
Seeds: Pollen cones are 10 to 12 inches (25 to 30 cm) long
Range: Northern Australia
Habitat: Limestone outcrops and in sandstone

WOOD'S CYCAD
(ENCEPHALARTOS WOODII)

Wood's cycad is a rare tree. It has a very thick trunk, and the tree is topped with a massive, umbrella-shaped crown of fronds. Each frond has two rows of thick, glossy leaves. Wood's cycad also has male and female plants. The male trees produce bright-orange pollen cones that can grow up to 4 feet (1.2 m) tall.

FUN FACT

In 1895, a botanist named John Medley Wood found a species of cycad he hadn't seen before. The plant was named Wood's cycad after him. In 1916, the last known specimen of Wood's cycad was collected from the wild. Today Wood's cycad lives on in botanical gardens and private collections. But there's still a chance there are specimens in the wild waiting to be found.

HOW TO SPOT

Height: Up to 20 feet (6 m)

Leaves: Glossy fronds, 6 to 9 feet (2 to 3 m) long

Seeds: Orange cones that grow up to 4 feet (1.2 m) long

Range: South Africa

Habitat: Steep, south-facing slopes

PINEAPPLE CYCAD
(LEPIDOZAMIA PEROFFSKYANA)

The pineapple cycad is one of the tallest cycads in the world, growing up to 23 feet (7 m) tall. Like many cycads, the pineapple cycad often looks like a palm tree. However, it is not closely related to the palm family—and it's not related to the plant that produces pineapples. The tree has a straight trunk and glossy, frond-like leaves that are brown and hairy at first before turning green and smooth.

Male pineapple cycads have long, spiral-shaped cones.

HOW TO SPOT

Height: Up to 23 feet (7 m)
Leaves: Fronds; 10 feet (3 m) long
Seeds: Male pollen cones and female seed cones; 7.5 to 9 inches (19 to 23 cm) long
Range: Eastern Australia
Habitat: Coastal forests

GINKGO *(GINKGO BILOBA)*

The ginkgo tree is the only species in its family. It's a large deciduous tree with ridged bark and green, fan-shaped leaves that turn bright yellow in the fall. Its seeds are enclosed in a fruit that looks like a cherry and turns yellow or orange when ripe.

HOW TO SPOT

Height: 40 to 70 feet (12 to 21 m)

Leaves: Notched, fan-shaped leaves; 3 inches (7.6 cm) long

Seeds: Yellow, berry-like seeds; 1 inch (2.5 cm) long

Range: China

Habitat: Grown in gardens or yards

FUN FACT

Scientists consider the gingko tree to be a living fossil. That's because it looks nearly the same as fossilized gingko trees from 270 million years ago.

MELINJO *(GNETUM GNEMON)*

As an evergreen tree, the melinjo's simple leaves are shiny and dark green year-round. Its trunk has swollen rings around it, marking where old branches once were. A few times a year, the melinjo produces catkins that later turn into clusters of small, oval-shaped seedpods. The seedpods are yellow at first before ripening and turning purple.

FUN FACT

The melinjo is a type of tree known as a gnetum. Gnetums belong to the gymnosperm family and grow only in tropical areas.

HOW TO SPOT

Height: Up to 50 feet (15 m)

Leaves: Broad, pointed ovals; 4 to 8 inches (10 to 20 cm) long

Seeds: Oval-shaped pods; 0.4 to 1.3 inch (1 to 3.3 cm) long

Range: Southeast Asia

Habitat: Tropical and subtropical rain forests

ALOE YUCCA *(YUCCA ALOIFOLIA)*

The aloe yucca is also known as the Spanish bayonet tree. That's because it has long, sword-like leaves that can be very sharp. The leaves grow along the tree's entire trunk. They turn brown and often fall off once they die. The greenest part of the tree is the very top, where a crown of green leaves grow in a sunburst shape. The aloe yucca's leathery seeds begin as large, white flowers.

FUN FACT

Angiosperms are the largest—and most diverse—group in the plant kingdom. This group includes 300,000 species of flowering plants.

HOW TO SPOT

Height: 5 to 20 feet (1.5 to 6 m)

Leaves: Pointed leaves; 2 feet (0.6 m) long

Fruit: Fleshy, green to black pods; 3 to 4 inches (7.6 to 10 cm) long

Range: Southern United States

Habitat: Coastal areas

YUCCA PLANTS

All yuccas are evergreen plants that have thick, blade-like leaves and that usually grow in warm, dry climates. There are 40 species of yucca plants worldwide, but most are shrubs.

JOSHUA TREE *(YUCCA BREVIFOLIA)*

The Joshua tree is the tallest of all yucca plants. This slow-growing tree has long, thin leaves that grow in spiky clusters at the ends of its branches. The trunk and branches of the tree are usually covered with a brown, shaggy layer of dead leaves. In the spring the Joshua tree produces pale-green flowers that transform into thick, egg-shaped seeds.

HOW TO SPOT

Height: 20 to 40 feet (6 to 12 m)

Leaves: Spike-like; 12 inches (30 cm) long

Fruit: Egg shaped; 2 to 4 inches (5 to 10 cm) long

Range: Southern California

Habitat: Desert

FUN FACT
The Joshua tree is a very rare tree. It grows only in one small region of the world—the Mojave Desert.

SAGUARO *(CARNEGIEA GIGANTEA)*

Because of its height and single trunk, the giant saguaro cactus is one of the few cacti to be considered a tree. Instead of being covered with bark, the saguaro's trunk is fleshy and green with many vertical grooves. Instead of leaves, this tree has long, closely packed spines on its trunk and branches. The saguaro produces white flowers that eventually become red, fleshy fruit. The fruit has sharp spines and can be filled with 2,000 seeds.

HOW TO SPOT

Height: 20 to 50 feet (6 to 15 m)
Leaves: Spines
Fruit: Red, oval fruit
Range: Southwestern United States and northeastern Mexico
Habitat: Deserts

FUN FACT

Flowers on a saguaro cactus open at night and bloom for less than one day.

QUIVER TREE
(ALOIDENDRON DICHOTOMUM)

The quiver tree is a large succulent that has a straight trunk and a rounded crown of repeatedly forked branches. The trunk and branches are usually coated with a thin layer of white powder that helps reflect the heat of the sun. The blue-green leaves are thick and rubbery, growing in rosettes from the tips of the branches.

HOW TO SPOT

Height: Up to 23 feet (7 m)

Leaves: Fleshy, blade-like leaves; 12 inches (30 cm) long

Fruit: Clusters of fleshy, yellow flowers that become oval-shaped fruit

Range: Southern Africa

Habitat: Rocky hillsides

FUN FACT

A succulent is a plant that has thick, fleshy leaves. The leaves store water for the plant. This is important because most succulents grow in dry climates.

AMERICAN BEECH
(FAGUS GRANDIFOLIA)

The American beech tree is a large deciduous tree. It is often recognizable in the winter because it hangs on to its pale-yellow leaves until spring. In the spring and summer, its leaves are dark green. American beech trees often have short trunks with their branches beginning near the ground. The trunk is covered in smooth, pale-gray bark.

FUN FACT

Beech nuts are rarely eaten by people. But they are an important food for bears, deer, birds, and mice.

HOW TO SPOT

Height: 50 to 70 feet (15 to 21 m)

Leaves: Pointed ovals with toothed edges; 4 inches (10 cm) long

Fruit: Round, spiky husks; 0.75 inches (1.9 cm) long

Range: Eastern United States

Habitat: Forests

ALGERIAN OAK
(QUERCUS CANARIENSIS)

The Algerian oak is considered a semievergreen because—depending on where it grows—most of its leaves stay green throughout the winter. It has large, green leaves that have eight to 12 shallow lobes. Like all oak trees, the Algerian oak produces drooping flowers called catkins in the spring that look like light-green tassels. Over the summer months, these catkins turn into small, brown acorns.

HOW TO SPOT

Height: 80 to 100 feet (24 to 30 m)

Leaves: Shallow, rounded lobes; 6 inches (15 cm) long

Fruit: Acorns; 1 inch (2.5 cm) long

Range: Northern Africa and southwestern Europe

Habitat: Forests

RED OAK *(QUERCUS RUBRA)*

One simple way to identify an oak tree is to examine its acorns. The red oak's acorns have a very shallow cap that is covered in small, smooth scales. The acorn's kernel is small and oval shaped. It turns from green to brown when it's ripe. The red oak can also be recognized by its pointed leaves—which turn red in the fall—and gray bark, which has shallow cracks and smooth ridges.

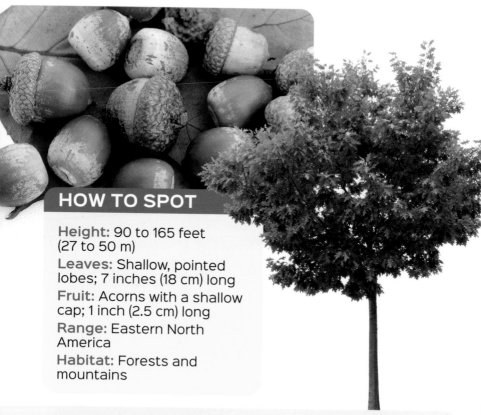

HOW TO SPOT

Height: 90 to 165 feet (27 to 50 m)

Leaves: Shallow, pointed lobes; 7 inches (18 cm) long

Fruit: Acorns with a shallow cap; 1 inch (2.5 cm) long

Range: Eastern North America

Habitat: Forests and mountains

OAKS AND ACORNS

All oak trees—no matter the species—produce acorns. But the acorns can look very different depending on the tree. All acorns have a cap, shell, and kernel. The caps can be smooth or rough and can cover a little of the shell or nearly all of it. The shape of the shell can also vary from nearly round to a long, oblong shape.

WHITE OAK *(QUERCUS ALBA)*

This tree is known for its short trunk and wide, spreading branches that often grow horizontal to the ground. It has some of the brightest fall foliage of all the oak trees, with its dark-green leaves turning a burgundy or orange-red color. White oaks produce oval-shaped acorns that grow in clusters of two to six. Each acorn has a bumpy cap.

FUN FACT

White oaks can produce acorns each year, and every four to seven years they produce an extra-large crop of acorns. Even then, whole acorns can be hard to find because they're a favorite food of squirrels and deer.

HOW TO SPOT

Height: 60 to 80 feet (18 to 24 m)

Leaves: Deep, rounded lobes; 6.5 inches (16.5 cm) long

Fruit: Acorns with a rough, scaly cap; 1 inch (2.5 cm) long

Range: Eastern North America

Habitat: Dry forests

AFRICAN TULIP TREE
(SPATHODEA CAMPANULATA)

The African tulip tree is a fast-growing evergreen tree that's found in tropical environments. It blooms one or more times each year. Its large, reddish-orange flowers look like wavy-edged tulips. The African tulip tree's compound leaves are green and look a little like ferns. Its seedpods split open when ripe to release as many as 500 winged seeds.

HOW TO SPOT

Height: 25 to 80 feet (7.5 to 24 m)

Leaves: Compound leaves that are 20 inches (50 cm); oval leaflets are 4 inches (10 cm)

Fruit: Pointed pods; 8 inches (20 cm) long

Range: Africa and southern United States

Habitat: Tropical forests

Simple leaf   Compound leaf

SIMPLE AND COMPOUND LEAVES

Most trees in the angiosperm family have simple leaves. This means each leaf stalk has a single leaf on it. Some trees, however, have compound leaves. This means that each leaf stalk has multiple small leaves called leaflets.

JACARANDA *(JACARANDA MIMOSIFOLIA)*

The jacaranda tree has crooked branches and a dense canopy. As a deciduous tree, it briefly loses its long, fern-like leaves in late winter. A short time later, in the spring, the jacaranda tree is full of purple flowers before growing a new set of leaves.

FUN FACT

The jacaranda tree is also called the fern tree. That's because of its feathery, fern-like leaves.

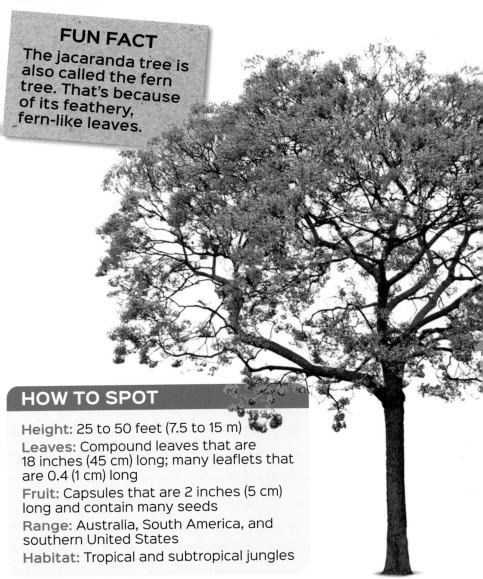

HOW TO SPOT

Height: 25 to 50 feet (7.5 to 15 m)

Leaves: Compound leaves that are 18 inches (45 cm) long; many leaflets that are 0.4 (1 cm) long

Fruit: Capsules that are 2 inches (5 cm) long and contain many seeds

Range: Australia, South America, and southern United States

Habitat: Tropical and subtropical jungles

EUROPEAN ALDER
(ALNUS GLUTINOSA)

The European alder is a medium-sized tree that often has multiple small trunks instead of one main trunk. Its bark is dark brown with many bumpy, horizontal stripes. The tree produces catkins that turn into small, flared cones—they look like tiny pine cones.

FUN FACT

The European alder tree often grows near water. It uses water to distribute its seeds, which float on the water's surface.

HOW TO SPOT

Height: 20 to 50 feet (6 to 15 m)

Leaves: Broad, fan-shaped leaves with both toothed and rounded edges

Fruit: Drooping catkins that turn to woody cones in female trees; 0.5 inch (1.2 cm) long

Range: Europe, Siberia, northeastern and midwestern United States, and Canada

Habitat: Near water or in moist soil

TURKISH HAZEL *(CORYLUS COLURNA)*

Most plants in the hazel family are shrubs, but the Turkish hazel is a tree. It's also the largest of all hazel plants. This tree grows in a broad cone shape with short branches that are often parallel to the ground. The tree's bark peels off in square flakes and strips. The Turkish hazel produces small hazelnuts that are enclosed in a thick shell and shaggy husk.

HOW TO SPOT

Height: 40 to 70 feet (12 to 21 m)

Leaves: Round with irregular teeth

Fruit: Nuts with a shaggy, paper-like husk; 0.5 inch (1.2 cm) long

Range: Southeastern Europe, southwestern Asia, and North America

Habitat: Open forests

PAPER BIRCH *(BETULA PAPYRIFERA)*

The most recognizable part of the paper birch is its white trunk covered in peeling bark. Young paper birches have smooth, reddish-brown bark. As the tree ages, the bark turns white and peels off in horizontal strips that look like paper. The layer underneath the peeling bark is smooth and pink. The paper birch's leaves are shaped like pointed ovals and have jagged-toothed edges.

HOW TO SPOT

Height: 70 to 80 feet (21 to 24 m)

Leaves: Toothed ovals with pointed tips; 3.5 inches (9 cm) long

Fruit: Flexible cones; 1 to 2 inches (2.5 to 5 cm) long

Range: North America

Habitat: Open forests

PAPER BIRCH USES

The paper birch is also called the canoe birch. The tree's flexible, waterproof bark was used by Native Americans such as the Ojibwe to build canoes. Because it's easy to burn and slow to rot, birch bark is considered one of the forest's most useful natural materials.

MANCHURIAN BIRCH
(BETULA PLATYPHYLLA)

The Manchurian birch is a small, single-trunked birch tree that thrives in cool, moist climates. It has thick, white bark that doesn't peel and dark-green, triangular leaves. Like many other birches, the Manchurian birch has thin branches that droop slightly at the ends. In the fall, the tree's female flowers turn into cone-like fruit full of tiny winged seeds.

FUN FACT
Manchuria is a region in northeast China. That is one area in which Manchurian birches grow.

HOW TO SPOT

Height: 30 to 40 feet (9 to 12 m)

Leaves: Broad, triangular leaves with toothed edges; 3.5 inches (9 cm) long

Fruit: Cone-like fruit

Range: China, Korea, Japan, and North America

Habitat: Uplands and clearings

BAOBAB *(ADANSONIA DIGITATA)*

The massive baobab tree can have a trunk that's more than 80 feet (24 m) around and covered in slippery, smooth, brown bark. The tree's unusual branch formation looks like twisted tree roots. When the baobab blooms, it has white flowers that hang straight down from the branches. Later, these flowers grow into velvety, yellow fruit.

HOW TO SPOT

Height: 16 to 65 feet (5 to 20 m)

Leaves: Fan-shaped compound leaves with five to nine leaflets

Fruit: Large, oval-shaped fruit; 8 inches (20 cm) long

Range: Africa, Madagascar, Southeast Asia, and Australia

Habitat: Sandy plains

DURIAN *(DURIO ZIBETHINUS)*

The durian tree is best known for its large fruit, called durian, that has an incredibly strong scent. The fruit is covered in short spines and can be green or yellowish brown. The tree's trunk has a buttressed base and peeling, reddish-brown bark. The tree blooms once or twice a year with feathery, white flowers.

FUN FACT

Because of its strong smell, durian fruit is banned from some public places in Southeast Asia—including the subway system in Singapore and hotels in Thailand, Japan, and Hong Kong. The durian has been described as smelling like onions, sulfur, and gym socks.

HOW TO SPOT

Height: Up to 147 feet (45 m) tall

Leaves: Pointed, oval leaves; 4 to 6 inches (10 to 15 cm) long

Fruit: Large, spiny, melon-shaped fruit; 6 to 10 inches (15 to 25 cm) long

Range: Southeast Asia and Australia

Habitat: Tropical jungles

CASHEW *(ANACARDIUM OCCIDENTALE)*

Though the cashew tree is most known for its nuts, this tropical tree also produces large, kidney-shaped fruit called cashew apples. The cashew nut grows on the very end of the apple in a tough shell. This tree can be recognized by its thick, leathery leaves that grow in a spiral around the branches.

HOW TO SPOT

Height: Up to 40 feet (12 m)

Leaves: Ovals with smooth edges; 4 inches (10 cm) long

Fruit: Kidney-shaped, fleshy fruit with a cashew nut on one end

Range: South America

Habitat: Coastal areas

MANGO *(MANGIFERA INDICA)*

The tropical mango tree has long, drooping leaves that grow outward from the tips of the branches into a rosette shape. When the mango tree blooms, it has pyramid-shaped clusters of red or yellow flowers. These flowers later develop into mangoes—the tree's fruit. There are many different varieties of mangoes.

HOW TO SPOT

Height: 30 to 100 feet (9 to 30 m)

Leaves: Narrow, blade-shaped leaves; 4 to 12.5 inches (10 to 32 cm) long

Fruit: Oval-shaped fruit ranging from 2.5 to 10 inches (6 to 25 cm) long

Range: Africa, South Asia, North America, and South America

Habitat: Coastal areas

FUN FACT

Depending on the cultivar, mangoes can be light green, dark green, yellow, yellow orange, red, or purplish red.

47

PISTACHIO *(PISTACIA VERA)*

The pistachio tree has a dense canopy of leaves that usually grow as three leaflets. Each leaflet is an irregular oval shape with a slightly wavy edge. The tree produces clusters of pistachios—small, oval-shaped nuts.

HOW TO SPOT

Height: Up to 30 feet (9 m) tall

Leaves: Compound leaves that are 8 inches (20 cm) long with three leaflets

Fruit: Pistachios; 0.75 inch (1.9 cm)

Range: Western and central Asia, southwestern United States, and the Mediterranean

Habitat: Dry, hot areas

STAGHORN SUMAC *(RHUS TYPHINA)*

The staghorn sumac is one of many varieties of sumac, but this tree is easy to recognize because of its upright clusters of berry-like seeds. The seed clusters often stay on the tree even after the leaves fall. The staghorn sumac's many small branches make it look like a very tall shrub, but it's actually considered to be a small tree. In the fall, the staghorn sumac's green leaves turn bright red.

FUN FACT
Staghorn sumac gets its name from having branches that look like deer antlers.

HOW TO SPOT

Height: 4 to 15 feet (1.2 to 4.5 m)

Leaves: Compound leaves that are 23 inches (58 cm) long with toothed-pointed oval leaflets

Fruit: Red berries in pointed clusters; clusters are 6 inches (15 cm) long

Range: Midwestern and northeastern United States

Habitat: Open areas

AMERICAN ELM *(ULMUS AMERICANA)*

American elms are large trees that create a lot of shade with their upward-spreading branches. The tree grows in a *V* shape. The American elm's green, almond-shaped leaves can be rough to the touch and feel a bit like sandpaper. They also have many straight veins that run from the leaf stems to the edges.

FUN FACT

Because of Dutch elm disease—a fungus that's spread by a certain type of beetle—the American elm is now rare in the wild. But it can still be seen in cities and neighborhoods.

HOW TO SPOT

Height: 80 to 100 feet (24 to 30 m)

Leaves: Pointed oval leaves with toothed edges; 6 inches (15 cm) long

Fruit: Round, papery fruit called samaras; 0.5 inch (1.2 cm) long

Range: North America

Habitat: Floodplains and stream banks

CHINESE ELM *(ULMUS PARVIFOLIA)*

The Chinese elm has a rounded crown and long, drooping branches. When this elm tree is mature, its bark will flake off and viewers can see orange, brown, green, gray, and even cream coloring underneath. In early fall, tiny red-green flowers will grow on the tree. Unlike the American elm, the Chinese elm has a strong resistance to Dutch elm disease.

HOW TO SPOT

Height: 40 to 60 feet (12 to 18 m)

Leaves: Dark-green leaves; 0.7 to 2 inches (1.8 to 5 cm) long

Seeds: Small seeds encased in a papery wing

Range: Korea, China, and Japan

Habitat: Various, such as prairies and meadows

WYCH ELM *(ULMUS GLABRA)*

The wych elm usually has a very short main trunk with multiple smaller trunks growing from it. When the tree is young, its bark is gray and very smooth. After 20 years, the bark becomes rough and cracked. The wych elm's rough leaves are larger than other elms but have a similar lopsided almond shape.

FUN FACT

Wych comes from an Old English word that means "bend." It was a name given to trees with flexible branches.

HOW TO SPOT

Height: 80 to 130 feet (24 to 40 m)

Leaves: Pointed ovals with toothed edges and a lopsided base; 5.5 inches (14 cm) long

Fruit: Oval paper samaras with notched ends; 0.75 inch (1.9 cm) long

Range: Europe, western Asia, and North America

Habitat: Rocky hillsides and forests

LUMBER AND ELM TREES

Elm trees were once an important source of lumber. That's because elm wood is a tough wood that has a tight, twisted grain and is water-resistant. Unlike other types of wood, it doesn't split easily, making it a good choice for furniture and building materials. However, since the 1930s, many elm trees have died as a result of Dutch elm disease, so elm wood is rarely used today.

JAPANESE ZELKOVA
(ZELKOVA SERRATA)

The Japanese zelkova is an elm tree. It has a broad, dense crown with twigs that droop slightly at the ends. Its bark is pale gray, but as the tree ages this bark sometimes flakes off to reveal orange-brown bark underneath. The Japanese zelkova produces tiny, green, berry-like seeds at the base of its leaves. In the fall, the Japanese zelkova's green leaves can turn a variety of bright colors—from red to orange to pink.

HOW TO SPOT

Height: Up to 100 feet (30 m) tall
Leaves: Narrow, pointed ovals with large teeth; 4 inches (10 cm) long
Fruit: Very small, green, berry-like fruit
Range: Japan, Korea, Taiwan, and North America
Habitat: Floodplains and stream banks

AVOCADO *(PERSEA AMERICANA)*

The avocado tree is most known for its large, green fruit—avocados. Each avocado begins as a tiny, green flower and can grow up to 7 inches (18 cm) long. The tree itself is medium sized, with many branches and a rounded crown. The leaves of the avocado tree are very thick and rigid, and they feel like leather.

HOW TO SPOT

Height: Up to 65 feet (20 m)

Leaves: Thick, leathery, almond-shaped leaves; 6 inches (15 cm) long

Fruit: Soft, green fruit with bumpy, green skin

Range: Central America and North America

Habitat: Tropical forests

FUN FACT

Another name for avocado is alligator pear.

BAY LAUREL *(LAURUS NOBILIS)*

The bay laurel is a small tree that often looks like a shrub due to its compact size and dense crown of short branches. Its leaves are thick and glossy green, and they release a pleasant scent when they're crushed. In the fall, the tree produces small, green berries that turn black as they ripen.

LAUREL TREE USES

The laurel family includes about 2,500 shrubs and trees. Many species in this family are known for their aromatic oils and fragrant leaves that are used in everything from perfumes to flavorings to spices. Bay leaves, cinnamon, and the original source for root beer flavoring all come from this large plant family.

HOW TO SPOT

Height: 20 to 30 feet (6 to 9 m)

Leaves: Smooth-edged, almond-shaped leaves that taper toward the base

Fruit: Small, round berries

Range: Mediterranean and North America

Habitat: Evergreen forests

55

CAMPHOR TREE
(CINNAMOMUM CAMPHORA)

The camphor tree is also known as the camphor laurel. It is an evergreen with wavy, oval-shaped leaves that release a strong fragrance when crushed. The tree loses its old leaves when the new leaves are ready to open. The camphor tree also produces strongly scented, white flowers that develop into tiny, black berries. The bark of the tree is grayish brown and deeply wrinkled in vertical grooves.

HOW TO SPOT

Height: Up to 67 feet (20.4 m)

Leaves: Pointed ovals with wavy edges; 5 inches (13 cm) long

Fruit: Small, berry-like fruit

Range: China, Japan, and the southern United States

Habitat: Moist soil

SASSAFRAS *(SASSAFRAS ALBIDUM)*

The sassafras tree has some unique characteristics, including irregular leaf shapes and eye-catching berries. The tree produces green leaves that can either be shaped like pointed ovals or can have three lobes. In the fall, the tree has small, dark-blue berries that hang from bright-red stems.

FUN FACT

The roots of the sassafras tree were once used to create the main flavor in root beer. Today most root beer is made with artificial flavors.

HOW TO SPOT

Height: 10 to 50 feet (3 to 15 m)

Leaves: Pointed oval leaves and lobed leaves; 6 inches (15 cm) long

Fruit: Dark-blue, berry-like fruit; 0.4 inch (1 cm) long

Range: Eastern United States

Habitat: Fields and along the edges of forests

SILVER WATTLE *(ACACIA DEALBATA)*

The silver wattle is an elegant evergreen tree that has feathery, fern-like leaves. It also produces round, yellow flowers that grow in clusters—the flowers look like fuzzy pom-poms. Like all acacia trees, the silver wattle is a part of the legume (bean) family, which is why its seedpods look like flattened green beans. As the seedpods ripen, they slowly turn brown.

HOW TO SPOT

Height: Up to 70 feet (21 m) tall

Leaves: Green, feather-like leaves made up of tiny leaflets; 4.75 inches (12 cm) long

Fruit: Thin, green pods; 3 inches (7.5 cm) long

Range: Southeastern Australia, Tasmania, and California

Habitat: Mountain gullies and stream banks

EASTERN REDBUD
(CERCIS CANADENSIS)

The eastern redbud is a small tree known for the tiny, purplish-pink flowers that line its branches in the spring. After the eastern redbud blooms, it grows green, heart-shaped leaves and bean-like seedpods that can be either green or red. Young eastern redbuds have gray bark with orange grooves. As the trees get older, the bark's texture changes and turns into small, gray scales with orange underneath.

FUN FACT
Some redbud species have bronze-colored leaves instead of green.

HOW TO SPOT

Height: Up to 32 feet (10 m)

Leaves: Heart shaped; 4 inches (10 cm) long

Fruit: Thin, flat pods; 2.5 inches (6 cm) long

Range: Western Asia, southeastern Europe, and the eastern United States

Habitat: Moist forests

HONEY LOCUST
(GLEDITSIA TRIACANTHOS)

Young honey locust trees have clusters of sharp spines that grow from their trunks and branches. As the tree gets older, these spines wear off and the gray-brown bark becomes rough with deep ridges. The honey locust's green leaves are made up of many small leaflets and turn yellow in the fall. The tree's twisted seedpods are dark reddish brown and often stay on the tree even in winter.

HOW TO SPOT

Height: 70 to 80 feet (21 to 24 m)

Leaves: The leaves are 6 inches (15 cm) long with 15 to 30 leaflets that are 1 inch (2.5 cm) long

Fruit: Flat, wrinkled seedpods; 8 inches (20 cm)

Range: North America

Habitat: Lowlands

PAGODA TREE
(STYPHNOLOBIUM JAPONICUM)

The pagoda tree is a large tree with dense, jagged branches. Its compound leaves are glossy green on top while underneath they have a duller appearance and fuzzy texture. The pagoda tree's seedpods look like strands of beads that hang down from the branches. The seedpods are yellowish green in the summer, turn black in the fall and winter, and don't fall off the tree until spring.

HOW TO SPOT

Height: 50 feet (15 m)

Leaves: Compound leaves are 10 inches (25 cm) long with seven or more pointed leaflets that are 2 inches (5 cm) long

Fruit: Yellowish-green pods; 5 inches (13 cm)

Range: China, Korea, Vietnam, and North America

Habitat: Forests and dry valleys

FUN FACT
The pagoda tree is also called the Chinese scholar tree.

AMERICAN LINDEN *(TILIA AMERICANA)*

The American linden is best recognized by its fine-toothed, heart-shaped leaves and by its fruit. In the spring, the American linden grows clusters of tiny flowers that become small, round seeds by the end of summer. Each seed cluster is attached to a special leaf called a bract that acts as a parachute when the seeds fall.

HOW TO SPOT

Height: 50 to 70 feet (15 to 21 m)

Leaves: Broad and heart shaped; 7 inches (18 cm)

Fruit: Clusters of small, round fruit

Range: Central and eastern North America

Habitat: Moist woods

SILVER LINDEN *(TILIA TOMENTOSA)*

The silver linden has two-toned leaves. The top is dark green, and the underside is a pale, silvery green. The leaves tend to be small and look like lopsided hearts with one side larger than the other. The silver linden produces small, pale-yellow flowers in the spring. These flowers turn to drooping clusters of nuts in the fall.

HOW TO SPOT

Height: 50 to 70 feet (15 to 21 m)

Leaves: Broad with a lopsided heart shape; 3.5 inches (9 cm)

Fruit: Clusters of round, tan nuts

Range: Southwestern Asia and southeastern Europe

Habitat: Deciduous and evergreen forests

FUN FACT

The silver linden's flowers have a strong scent that pollinators love. Honeybees often prefer its flowers to any others in the area.

SOUTHERN MAGNOLIA
(MAGNOLIA GRANDIFLORA)

The southern magnolia is an evergreen tree that has huge, white flowers every spring and sometimes in the fall. The southern magnolia's leaves are very thick and stiff with a leathery texture. The top sides of the leaves are a glossy dark green, and the undersides are covered in rust-colored hairs, making them appear brown. The tree's gray-brown bark is smooth but develops flaky scales as it gets older.

HOW TO SPOT

Height: 60 to 80 feet (18 to 24 m)
Leaves: Oval; 6 inches (15 cm)
Fruit: Egg-shaped, red cones; 3 to 4 inches (7.5 to 10 cm)
Range: Southeastern United States
Habitat: Moist forests and lowlands

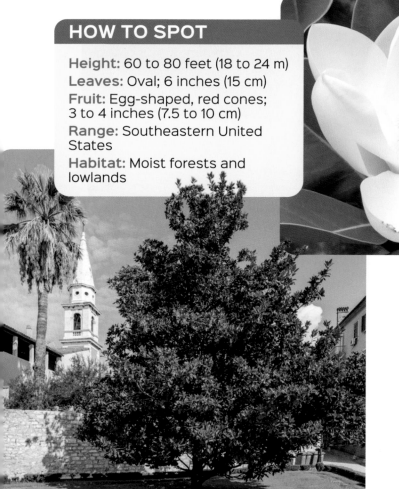

TULIP TREE *(LIRIODENDRON TULIPIFERA)*

The green-and-orange flowers that appear on its branches each summer give the tulip tree its name. The flowers sit upright on the twigs and later in the year turn into a seed cluster that looks like a cone. The tulip tree's four-lobed leaves are unique, with two lobes at the end of the leaf and one on each side.

FUN FACT

George Washington planted tulip trees that can still be seen today. They're 140 feet (43 m) tall!

HOW TO SPOT

Height: 50 to 200 feet (15 to 60 m)

Leaves: Unique four-lobed shape; 5 inches (13 cm) long

Fruit: Cone-like seed cluster

Range: Eastern United States

Habitat: Mixed deciduous forests and fields

KAPOK TREE *(CEIBA PENTANDRA)*

The tall, tropical kapok tree has an incredibly thick trunk. It can be up to 9 feet (3 m) or more in diameter. This massive trunk often has buttresses at its base, which are extra roots that grow from the trunk to help support the tree. The kapok tree has large, fan-shaped compound leaves and produces white-and-pink flowers. Later, these flowers become woody seed capsules. The seed capsules contain cotton-like fibers, called kapok, and around 200 tiny seeds.

HOW TO SPOT

Height: 75 to 125 feet (23 to 38 m)

Leaves: Compound leaves with five to nine leaflets; 3 to 7 inches (7.5 to 18 cm)

Fruit: Woody, oval-shaped capsules; 3 to 6 inches (7.5 to 15 cm)

Range: Southern North America, Central America, South America, and Africa

Habitat: Tropical jungles

KAPOK FIBERS

The kapok tree was an important source of kapok—a natural fiber that had many uses before artificial fibers were invented. Kapok is the soft, cotton-like substance that's found inside the tree's seedpods. It's naturally water-resistant and has a texture similar to soft, fluffy feathers. In the past, kapok was commonly used as stuffing for life jackets, mattresses, pillows, and furniture upholstery.

SEA HIBISCUS *(HIBISCUS TILIACEUS)*

The sea hibiscus's funnel-shaped, yellow flowers last only one day. But since the tree is an evergreen, it keeps its green, heart-shaped leaves year-round. The leaves are bright green on top and dull green and fuzzy on the underside. The sea hibiscus has multiple thin trunks and branches that often spiral and intertwine.

FUN FACT
In Asia, the sea hibiscus is harvested for its fiber, which is used to make rope.

HOW TO SPOT

Height: 12 to 25 feet (3.7 to 7.5 m)

Leaves: Heart shaped; 5 to 10 inches (13 to 25 cm)

Fruit: Small, brown, oval capsules

Range: Africa, Asia, Australia, North America, and the Pacific Islands

Habitat: Coasts

TALL-STILTED MANGROVE
(RHIZOPHORA APICULATA)

The tall-stilted mangrove is a tropical tree. It is usually found in intertidal areas along coasts, which are areas where the tide comes in and out, or at the mouths of rivers. It can have one or more thin trunks and is supported by a network of prop roots that grow from the tree's trunk to the water. The tall-stilted mangrove has thick, leathery leaves. Once a year the tree produces fuzzy, yellow-and-white flowers.

HOW TO SPOT

Height: 100 to 130 feet (30 to 40 m)

Leaves: Oval shaped with smooth edges

Fruit: Pear-shaped fruit

Range: Southeast Asia, Micronesia, and Australia

Habitat: Intertidal wetlands along tropical and subtropical coasts

MANGROVE FORESTS

Mangrove forests that thrive in coastal areas create important ecosystems. With their vast networks of underwater roots, mangroves provide shelter for young fish and crustaceans. Mangroves also provide habitat for larger subtropical and tropical animals such as birds, manatees, crocodiles, and panthers. These forests also help protect against erosion and flooding by acting as a buffer from the wind and waves.

RED MANGROVE
(RHIZOPHORA MANGLE)

The red mangrove is a tropical tree that grows in thick groves along coasts directly in salt water. It has a thick tangle of arch-shaped roots that grow both above and below water. Its leaves are evergreen and grow in clusters near the twig tips. The red mangrove's small, yellow flowers become leathery fruit containing the tree's seeds.

FUN FACT

The red mangrove is sometimes called the walking tree. That's because its roots make it appear as if it's walking on the surface of the water.

HOW TO SPOT

Height: Up to 80 feet (24 m)

Leaves: Wedge shaped; 2 to 6 inches (5 to 15 cm)

Fruit: Brown fruit; 1 to 1.5 inch (2.5 to 4 cm)

Range: Southern United States

Habitat: Intertidal wetlands along tropical and subtropical coasts

AMUR MAPLE *(ACER GINNALA)*

The Amur maple is a small maple tree that can often look like a lilac bush or other shrub. It usually has multiple small trunks and a rounded canopy. In the fall, Amur maples can turn yellow or bright red and produce winged seeds called samaras. The samaras are green at first but turn tan and papery when they're ripe.

Samaras

HOW TO SPOT

Height: 15 to 20 feet (4.5 to 6 m)

Leaves: Three-lobed leaves with toothed edges; 2.5 inches (6 cm) long

Fruit: Pairs of winged samaras; 1 inch (2.5 cm) long

Range: Asia, southeastern Europe, and North America

Habitat: Forests

RED MAPLE *(ACER RUBRUM)*

The red maple gets its name from its vibrant fall color. Most red maples have leaves that turn bright shades of yellow, orange, red, and sometimes maroon. In the summer, the leaves are green on top and pale, whitish green on the underside. These leaves—like all maple leaves—are lobed, which means the edges of the leaf go in and out, forming finger-like sections.

HOW TO SPOT

Height: 40 to 70 feet (12 to 21 m)

Leaves: Three triangular lobes with toothed edges; 4 inches (10 cm) long

Fruit: Pairs of winged samaras; 1 inch (2.5 cm) long

Range: North America

Habitat: Wooded swamps and near ponds

FUN FACT
The red maple has the longest north-to-south range of any tree in the eastern United States. It can be found from Newfoundland and Labrador all the way to southern Florida.

SILVER MAPLE *(ACER SACCHARINUM)*

The jagged leaves and pointed lobes of the silver maple only look sharp—they're actually as soft as any other maple leaf. The silver maple's green leaves have a silvery-green underside. These leaves turn yellow in the fall. The tree's bark is gray and smooth when the tree is young and begins to get rough and flaky as the tree ages, showing the brown wood underneath.

HOW TO SPOT

Height: 70 feet (21 m)

Leaves: Five-lobed with toothed edges

Fruit: Winged samaras; 1 to 2 inches (2.5 to 5 cm)

Range: Eastern North America

Habitat: Riverbanks

SUGAR MAPLE *(ACER SACCHARUM)*

The sugar maple is a large tree with a straight, grooved trunk and a dense canopy of branches and twigs. Like many maples, its leaves turn bright colors in the fall. The leaves on the sugar maple's outer branches change color first. So, the center of the tree still has glowing green leaves while the outside leaves are shades of gold, orange, and red.

FUN FACT

All maple trees produce sap that can be made into maple syrup. But the sugar maple has the highest amount of sugar in its sap.

HOW TO SPOT

Height: 40 to 80 feet (12 to 24 m)

Leaves: Five lobed; 5 inches (13 cm) long

Fruit: Winged samaras; 1 inch (2.5 cm) long

Range: Eastern North America

Habitat: Forests

MAPLE SAP AND SYRUP

In February or March, a combination of freezing nights and above-freezing days causes the sap in maple trees to run. The sap is collected by drilling a hole and inserting a metal spout into the tree. The sap drips very slowly. It takes about 40 gallons (150 L) of sap to make just 1 gallon (3.8 L) of maple syrup.

COMMON FIG *(FICUS CARICA)*

The common fig is a subtropical tree that can sometimes look like a shrub because of its many branches. Its network of spreading branches can be up to 150 feet (45 m) wide! This tree is most known for its fruit. Figs are a plump, teardrop-shaped fruit that can be green, yellow, or deep purple when ripe. The common fig's leaves are very large despite the tree's small size.

HOW TO SPOT

Height: 10 to 20 feet (3 to 6 m)

Leaves: Three to five rounded lobes with toothed edges; up to 9 inches (23 cm) long

Fruit: Teardrop-shaped fruit called a fig; 3 inches (7.5 cm) long

Range: Middle East, Western Asia, eastern Mediterranean, and North America

Habitat: Open areas

WHITE MULBERRY *(MORUS ALBA)*

The white mulberry is one of a few mulberry trees grown for its cylinder-shaped fruit, which starts out small and green. As the berries ripen, they turn white, red, or black. The large leaves of the white mulberry can be different shapes even on the same tree. It's common to see toothed leaves without any lobes, leaves with two lobes, and leaves with three lobes.

HOW TO SPOT

Height: 30 to 60 feet (9 to 18 m)

Leaves: Can be lobed or unlobed; 3 to 10 inches (7.5 to 25 cm) long

Fruit: Cluster-like berries; berry is 0.75 inch (1.9 cm) long

Range: Australia, China, Europe, Middle East, and North America

Habitat: Edges of forests

FUN FACT

The white mulberry was introduced to the United States by British colonists who wanted to use the tree as food for silkworms.

BLUE GUM EUCALYPTUS
(EUCALYPTUS GLOBULUS)

Unlike most trees, eucalyptus trees will produce differently shaped leaves as the tree gets older. On young blue gum eucalyptus trees, the leaves are short and broad—about 4 inches (10 cm) long. As the tree ages, it will start producing leaves that are thin and up to 10 inches (25 cm) long. The blue gum eucalyptus also has white flowers that look like powder puffs.

FUN FACT
There are around 800 species of trees in the *Eucalyptus* family.

HOW TO SPOT

Height: 80 feet (24 m)

Leaves: Thin and blade shaped; 10 inches (25 cm)

Fruit: Bowl-shaped capsules; 1 inch (2.5 cm)

Range: Southern Australia, Tasmania, and North America

Habitat: Coasts and hillsides

WEEPING BOTTLEBRUSH
(CALLISTEMON VIMINALIS)

The weeping bottlebrush is part of a larger family of bottlebrush trees. It gets its name from its long, red flower spikes. These flowers can bloom in the spring, fall, or throughout the year. They eventually develop into clusters of tiny, round fruit containing the tree's seeds. The branches of the weeping bottlebrush are flexible and droop slightly downward.

FUN FACT
All species of bottlebrush trees are native to Australia.

HOW TO SPOT

Height: Up to 59 feet (18 m)

Leaves: Narrow, pointed leaves; 1.5 to 2.7 inches (4 to 7 cm) long

Fruit: Small spike clusters of round fruit

Range: Eastern Australia

Habitat: Riverbanks, stream banks, and coastal plains

77

COMMON GUAVA *(PSIDIUM GUAJAVA)*

The common guava is a small tropical tree with flaky, copper-colored bark and green, oval-shaped leaves. Twice a year, the tree produces guavas. On the outside, the round, fleshy fruit can range from yellow to green. The guavas are filled with many small, hard seeds. Some guava varieties have soft, chewable seeds.

FUN FACT

In some parts of the world, the common guava is considered a weed because it spreads quickly in the wild and forms dense thickets.

HOW TO SPOT

Height: Up to 33 feet (10 m)

Leaves: Pointed ovals; 2.7 to 6 inches (7 to 15 cm) long

Fruit: Round or pear-shaped fruit filled with small seeds; 2 to 4 inches (5 to 10 cm) long

Range: Caribbean, Central America, southern North America, Asia, Africa, southern Europe, and the Pacific Islands

Habitat: Fields and coastal areas

CHILEAN MYRTLE *(LUMA APICULATA)*

The Chilean myrtle is a small evergreen tree. It usually has many small trunks and can sometimes look like a shrub. The leaves are glossy and green. They give off a pleasant scent when broken or crushed. Leaves on young Chilean myrtles usually are bronze colored at their tips. This tree also produces white, four-petaled flowers that later become berries. The berries are purplish black when ripe.

HOW TO SPOT

Height: Up to 60 feet (18 m)

Leaves: Smooth, pointed ovals; 1 inch (2.5 cm) long

Fruit: Round, purple-black berries

Range: Argentina, Chile, and North America

Habitat: Forests

WHITE ASH *(FRAXINUS AMERICANA)*

The white ash is the most common ash tree. It has a tall, straight trunk that has diamond-shaped grooves in the gray bark. It also has long compound leaves that each have about five to 13 leaflets. In the fall, these leaves can turn yellow, maroon, or dark purple. Each of the white ash's seeds has a single paper wing that helps transport them when they're ripe.

FUN FACT

The white ash is the most common ash tree used to make baseball bats.

HOW TO SPOT

Height: 50 to 80 feet (15 to 24 m)

Leaves: Leaves are about 12 inches (30 cm) long; leaflets are 4 inches (10 cm) long

Fruit: Winged fruit; 1.5 inches (4 cm) long

Range: Central and eastern North America

Habitat: Open areas and near streams

EUROPEAN ASH *(FRAXINUS EXCELSIOR)*

The large European ash is a common sight in forests throughout Europe, but it can sometimes be seen in the northeastern United States. Rather than turning a different color in autumn, the European ash's leaves fall from the tree while they're still green. The compound leaves are made up of nine to 11 toothed leaflets. The European ash has winged seeds that hang down in thick clusters from the branches.

HOW TO SPOT

Height: 60 to 80 feet (18 to 24 m)

Leaves: Leaves are 11 inches (28 cm) long; leaflets are 3 inches (7.5 cm) long

Fruit: Winged fruit; 1.5 inches (4 cm) long

Range: Europe and North America

Habitat: Forests

FLOWERING ASH *(FRAXINUS ORNUS)*

In the spring, the flowering ash produces cone-shaped clusters of fragrant, white flowers. Over the summer, these flowers turn into clusters of narrow winged seeds. The seeds are green at first but turn brown as they ripen. The bark of the flowering ash remains smooth and gray no matter the age of the tree.

HOW TO SPOT

Height: 30 to 50 feet (9 to 15 m)

Leaves: Compound leaves that are 7 inches (18 cm) long; five to nine leaflets that are 3 inches (7.5 cm) long

Fruit: Narrow winged fruit; 1.5 inches (4 cm) long

Range: Southwestern Asia, southern Europe, and North America

Habitat: Woods on hillsides

ASH TREE WOOD

There are around 55 species of ash trees around the world. The wood from ash trees is known for being very strong. It's been used in many things, including musical instruments, furniture, and tool handles.

EUROPEAN OLIVE *(OLEA EUROPAEA)*

The gnarled, dimpled trunk of the European olive tree gives it a unique appearance. It has a very dense crown of narrow leaves that are thick and feel like leather. As an evergreen, the European olive keeps its leaves year-round. Its small, yellow-white flowers become round olives that turn from green to black as they ripen.

HOW TO SPOT

Height: 20 to 30 feet (6 to 9 m)

Leaves: Narrow, pointed ovals; 2 inches (5 cm) long

Fruit: Oval-shaped olives; 1.5 inches (4 cm) long

Range: Southwestern Asia, southern Europe, and California

Habitat: Sunny open areas

FUN FACT
The olives that people eat such as Kalamata, Niçoise, and others are cultivars of the European olive tree.

AFRICAN OIL PALM
(ELAEIS GUINEENSIS)

The African oil palm is a rain forest tree. It is a tall tree with a ringed trunk and a large, fan-shaped crown. The crown is made up of very long, feather-like leaves that can have more than 100 pairs of leaflets. The African oil palm produces huge clusters of small, oval-shaped fruit. There can be 200 to 300 fruit in a single cluster.

HOW TO SPOT

Height: 27 to 65 feet (8 to 20 m)

Leaves: Compound leaves; 11 to 16 feet (3 to 5 m) long

Fruit: Oval-shaped, fleshy fruit; 1.3 inches (3 cm)

Range: West Africa and South America

Habitat: Tropical rain forests

CABBAGE PALMETTO
(SABAL PALMETTO)

Like most palms, the fan-shaped leaves of the cabbage palmetto are very large—they can be up to 6 feet (1.8 m) long. The leaves curl up slightly at the end and are made up of many stiff leaflets. Most of the tree's leaves fall off once they die, but some stay attached, forming a thick mat below the crown. When the cabbage palmetto is young, its gray bark has thin, vertical grooves, but the bark gets smoother with age.

HOW TO SPOT

Height: 40 to 50 feet (12 to 15 m)
Leaves: Fan-shaped compound leaves; 4 to 6 feet (1.2 to 1.8 m) long
Fruit: Small, berry-like fruit
Range: Southeastern United States
Habitat: Coasts

FUN FACT
The cabbage palmetto is able to withstand cooler temperatures than other North American palm trees. It can be found as far north as South Carolina.

COCONUT PALM *(COCOS NUCIFERA)*

The coconut palm has a round crown of large, feather-like leaves and a tall, smooth trunk that often bends or leans. The tree is best known for what it produces: coconuts! Coconuts grow in clusters, beginning as green and turning brown as they ripen. The coconut seed itself is enclosed within a thick husk.

FUN FACT

The strong fiber from the brown husk of coconuts can be used to make rope.

HOW TO SPOT

Height: 50 to 100 feet (15 to 30 m)
Leaves: Feather-shaped compound leaves; 22 feet (7 m) long
Fruit: Coconuts; 5 to 10 inches (13 to 25 cm) long
Range: Africa, Asia, Caribbean, southern North America, and the Pacific Islands
Habitat: Coasts

CALIFORNIA FAN PALM
(WASHINGTONIA FILIFERA)

The California fan palm is one of the largest palm trees. Unless they're trimmed off, the California fan palm's dead leaves build in layers along its trunk to form a dense thatch. The tree's berry-like fruit is easy to spot. It grows in clusters on stalks that are even longer than the leaves.

HOW TO SPOT

Height: Often 40 to 50 feet (12 to 15 m)

Leaves: Fan-shaped compound leaves; 6 feet (1.8 m) long

Fruit: Berry-like fruit; 0.4 inches (1 cm) long

Range: Northwestern Mexico and southeastern United States

Habitat: Deserts or dry soils

PALMS AND THEIR LEAVES

All palm trees have compound leaves that can be one of two types: pinnate or palmate. Pinnate leaves look like large feathers, with leaflets growing on opposite sides of a central stem. One palm with pinnate leaves is the cabbage palmetto.

Palmate leaves look like a hand (or fan) with leaflets growing out from a central point at the bottom of the stem. One palm with palmate leaves is the California fan palm.

CHINESE TALLOW TREE
(TRIADICA SEBIFERA)

The small Chinese tallow tree has diamond-shaped leaves that hang straight down from the branches. Its flowers grow in short, upright spikes and become brown three-section seed capsules in the fall and winter. When the seed capsules ripen to reveal the white seeds inside, they look like popcorn, which is why the Chinese tallow tree is also called the popcorn tree.

FUN FACT
In Asia, the waxy seed coating of the Chinese tallow tree is used to make candles and soap.

HOW TO SPOT

Height: 25 to 40 feet (7.5 to 12 m)

Leaves: Oval or diamond shaped; 3 inches (7.5 cm) long

Fruit: Small capsules

Range: China, Japan, Korea, and southeastern North America

Habitat: Wetlands and coastal plains

RUBBER TREE *(FICUS ELASTICA)*

The rubber tree has a dense canopy of stiff, glossy leaves that are red when they first open but quickly turn dark green. The rubber tree often has multiple small trunks and can also develop unique roots. The roots can form on the branches and hang from the tree before taking root in the ground.

HOW TO SPOT

Height: 35 to 45 feet (11 to 14 m)
Leaves: Pointed ovals; 3.5 to 12 inches (9 to 30 cm) long
Fruit: Small, egg-shaped, fleshy fruit
Range: Asia
Habitat: Jungles

ALMOND *(PRUNUS DULCIS)*

The almond tree's edible nut—almonds—are hidden inside a fleshy fruit that looks like a green peach. The fruit is thin and sour. In the spring, the almond tree is covered in bright-pink flowers before growing its narrow, dark-green leaves. The tree bark is dark gray and develops scales as the tree gets older.

HOW TO SPOT

Height: 20 feet (6 m)

Leaves: Narrow, blade-like leaves with fine-toothed edges; 4.25 inches (11 cm) long

Fruit: Almonds

Range: North Africa, central and southwestern Asia, southern Europe, and southwestern United States

Habitat: Forests and dry hillsides

COMMON APPLE *(MALUS DOMESTICA)*

Though species of the common apple are usually grown in orchards, they also grow in forests or open fields. In the spring, apple trees are covered with flowers that bloom as their leaves open. The round fruit of the apple tree can be many sizes and colors. Colors can range from vivid green, bold red, pink, yellow, brown, or a combination of these colors. Each apple contains many tiny, brown seeds at the center.

HOW TO SPOT

Height: 30 feet (9 m)

Leaves: Pointed ovals with toothed edges; 4.7 inches (12 cm) long

Fruit: Apples; variable in size

Range: Temperate climates all over the world, such as Asia, Europe, North America, Australia, and New Zealand

Habitat: Forests, open areas, and orchards

FUN FACT

All edible apples (other than crab apples) are different cultivars of the common apple tree.

ROSE FAMILY

Most fruit trees that grow in cooler climates—such as apple, plum, pear, and cherry—belong to the rose family. This is a plant family that includes more than 3,000 species. Most trees in the rose family have symmetrical, five-petaled flowers and fleshy fruit. All of these trees began as wild species, but most are now grown commercially at orchards or farms.

WILD SWEET CHERRY
(PRUNUS AVIUM)

The small wild sweet cherry tree often has a short, crooked trunk that's covered with glossy, reddish-brown bark. The bark tends to peel in horizontal strips. In the spring, the wild sweet cherry tree blooms with clusters of white flowers that are light pink in the center. By midsummer, the tree produces small cherries that are red or black when ripe.

HOW TO SPOT

Height: 30 to 50 feet (9 to 15 m)

Leaves: Pointed ovals with toothed edges; 4.5 inches (11 cm) long

Fruit: Small cherries

Range: Europe and North America

Habitat: Forests and hedgerows

PEACH *(PRUNUS PERSICA)*

The peach tree is known for its sweet, edible fruit. Peaches have soft, velvety skin and contain the tree's seed within their tough center pits. The peach tree's long, slender leaves are slightly curved and droop from the branches. Before the leaves open in the spring, the tree blooms with fragrant pink, white, or red flowers.

FUN FACT

A peach's large center pit can also be called a stone. Fruits that have similar seed capsules are called stone fruits.

HOW TO SPOT

Height: Under 20 feet (6 m) tall

Leaves: Narrow, pointed ovals; 6 inches (15 cm) long

Fruit/Seeds: Peaches containing a tough "stone," with a white seed inside

Range: Asia and North America

Habitat: Mountains, forests, and orchards

GRAPEFRUIT *(CITRUS PARADISI)*

The grapefruit tree has a broader crown than most other citrus trees, with its dense canopy often reaching the ground. Its pointed oval leaves are shiny on one side and dull on the other. Depending on the climate, the grapefruit tree blooms at least once a year with small clusters of white flowers. These flowers eventually grow into large, round fruit: grapefruit. Grapefruit can range in color from yellow to orange or shades of pink.

HOW TO SPOT

Height: 15 to 20 feet (4.5 to 6 m)

Leaves: Broad, pointed ovals; 6 inches (15 cm) long

Fruit: Grapefruit; 6 inches (15 cm)

Range: Caribbean, Mexico, southeastern and southwestern United States, and South America

Habitat: Groves and coastal areas

LEMON *(CITRUS LIMON)*

The lemon tree is popular throughout the world for its sour citrus fruit. Lemons vary in size and shape, but all are yellow and slightly oval shaped. The lemon tree's branches have long, sharp thorns, and its glossy leaves are pointed ovals with smooth or wavy edges.

FUN FACT

Citrus trees—orange, lemon, lime, tangerine, pomelo, and grapefruit—belong to the rue family.

HOW TO SPOT

Height: 10 to 20 feet (3 to 6 m)

Leaves: Pointed ovals; 6 inches (15 cm) long

Fruit: Lemons; 4 inches (10 cm) long

Range: Northern Africa, southern Europe, Middle East, southern North America, and Central America

Habitat: Groves and coastal areas

MEXICAN LIME *(CITRUS AURANTIFOLIA)*

Mexican lime trees are small, twisted trees that can often look like shrubs. Similar to lemon trees, they have long, sharp spines on their branches. After producing white flowers, the tree develops limes that appear alone, in pairs, or sometimes in clusters. The limes are a shiny green when they're young but turn a pale yellow-green when they're ripe.

HOW TO SPOT

Height: 6.5 to 13 feet (2 to 4 m)

Leaves: Broad, pointed ovals; 2 to 3 inches (5 to 7.5 cm) long

Fruit: Limes; 1 to 2 inches (2.5 to 5 cm) long

Range: North Africa, Asia, Caribbean, southern Europe, and southern North America

Habitat: Groves and coastal areas

FUN FACT

The Mexican lime is also known as the key lime. The tree's fruit is one of the ingredients in key lime pie.

SWEET ORANGE *(CITRUS SINENSIS)*

All edible varieties of oranges are variations of the sweet orange. This medium-sized citrus tree has dense, slender branches that angle upward. It can produce fragrant, white flowers throughout the year that later develop into oranges. Depending on the specific species, ripe oranges can be orange, orange red, or yellow.

HOW TO SPOT

Height: Up to 50 feet (15 m)

Leaves: Narrow and broad pointed oval leaves; 6 inches (15 cm) long

Fruit: Oranges; 2.5 to 3.7 inches (6 to 9 cm) long

Range: Caribbean, South Africa, Asia, Australia, southern Europe, southeastern and southwestern North America, and South America

Habitat: Groves and coastal areas

SWEET ORANGE HISTORY

The sweet orange species likely first grew in southern China, northeastern India, or southeastern Asia. It was probably brought to Europe between 1450 and 1500. In the mid-1500s, the Spanish brought this tree to Mexico and South America. These trees are no longer found in the wild today. Any orange trees that appear in the wild "escaped" from nearby groves.

SHAGBARK HICKORY *(CARYA OVATA)*

The shagbark hickory gets its name from the shaggy, ragged bark that covers its trunk. The tree has compound leaves of five leaflets with the top center leaflet being the largest. In autumn, the shagbark hickory produces large, green fruit. Inside the fruit is the ribbed hickory nut that splits into four sections.

FUN FACT

The shagbark hickory doesn't develop its telltale shaggy bark until it's at least 25 years old.

In the fall, shagbark hickory leaves turn yellow.

HOW TO SPOT

Height: 60 to 90 feet (18 to 27 m)

Leaves: Compound leaves; 11 inches (28 cm) long

Fruit: Round fruit with a nut inside; fruit is 1.75 inches (4 cm) long

Range: Eastern North America

Habitat: Forests and valleys

PECAN *(CARYA ILLINOINENSIS)*

The pecan is a very big tree with long, feathery leaves made up of curved leaflets. The tree begins with a single trunk that forks into multiple trunks not far off the ground. Pecan trees produce large, green, oval-shaped fruit. Inside a ripe fruit is a thin, black husk covering the pecan—a striped nut.

HOW TO SPOT

Height: 100 to 120 feet (30 to 36 m)
Leaves: 16 inches (40 cm) long
Fruit: Oblong fruit; 1.5 inches (4 cm) long
Range: Southern United States
Habitat: Moist forests and valleys

BLACK WALNUT *(JUGLANS NIGRA)*

The black walnut tree has a tall, straight trunk with gray bark that's wrinkled in a diamond pattern. It's common for several of the black walnut's main branches to be horizontal to the ground. In autumn, the tree produces round, green fruit. Inside the fruit is a walnut—an irregular-shaped nut with deep grooves.

HOW TO SPOT

Height: 50 to 75 feet (15 to 23 m)

Leaves: Compound leaves; 18 inches (46 cm) long

Fruit: Round fruit; 2 inches (5 cm) long

Range: Central and eastern United States

Habitat: Forests and slopes

TREE TOXIN

Black walnut trees have a special strategy for keeping away plants that might compete for food, water, and sunlight. The roots of black walnut trees release a chemical called juglone that's toxic to the roots of other plants such as pine, rhododendron, apple trees, and arborvitae.

ENGLISH WALNUT *(JUGLANS REGIA)*

The English walnut's compound leaves are made up of seven to nine leaflets including one large leaflet on the very end. The leaflets are rounded ovals with smooth edges. Walnuts—popular, edible nuts—are inside the tree's round, green fruit. In autumn, the fruit's glossy husk splits partly open, revealing the walnut inside.

HOW TO SPOT

Height: 40 to 60 feet (12 to 18 m)

Leaves: Compound leaves; 12 inches (30 cm) long

Fruit: Round, glossy fruit; 2.5 inches (6 cm) long

Range: Central and eastern United States, southeastern Europe, India, and China

Habitat: Stream banks and valleys

FUN FACT
Most walnuts sold in stores today are from English walnut trees.

QUAKING ASPEN
(POPULUS TREMULOIDES)

The round, fluttering leaves on the quaking aspen give the tree its name. The leaf stems are long and flat, which makes the leaves tremble even in a light breeze. The bark of young quaking aspens can be white, green, gray, or bronze. The bark of older trees is usually white with dark scars or horizontal ridges.

FUN FACT
Groves of aspen trees can often be one plant. Many trunks will share the same root system.

HOW TO SPOT

Height: 20 to 50 feet (6 to 15 m)

Leaves: Round or diamond shaped with pointed ends and toothed edges

Seeds: Tassel-like catkins; 4 inches (10 cm) long

Range: Northern North America

Habitat: Forests, fields, and roadsides

Quaking aspens turn yellow gold in the fall.

EASTERN COTTONWOOD
(POPULUS DELTOIDES)

Because eastern cottonwoods are fast-growing trees, even young trees can look ancient with massive, twisted trunks and wide, crooked branches. The branches are covered with triangular or heart-shaped leaves that extend from the twigs in many different directions. When the eastern cottonwood's catkins are ripe in early summer, they release fluffy, cotton-like seeds that float through the air.

HOW TO SPOT

Height: 60 to 90 feet (18 to 27 m)

Leaves: Triangular or heart-shaped leaves; 5 inches (13 cm) long

Seeds: Tassel-like catkins; 4 inches (10 cm) long

Range: Central North America

Habitat: Riverbanks, stream banks, and moist forests

WHITE POPLAR *(POPULUS ALBA)*

The leaves of the white poplar can be a few different shapes. Sometimes they're lobed like maple leaves, and sometimes they're triangular with wavy edges. No matter the shape, the leaves are always dark green on top and have fuzzy, white undersides covered with tiny hairs. The scarred, white trunk can look similar to birches and aspens.

HOW TO SPOT

Height: 60 to 80 feet (18 to 24 m)
Leaves: Three lobed or triangular with wavy edges
Seeds: Tassel-like catkins; 4 inches (10 cm) long
Range: North Africa, central and western Asia, Europe, and North America
Habitat: Forests

FUN FACT

Many North American trees that look like white poplars are not. Most are gray poplars. This is a hybrid of the white poplar and European aspen. Gray poplars have more rounded leaves that have a thinner layer of fuzz on their undersides.

WEEPING WILLOW *(SALIX BABYLONICA)*

The unique look of the weeping willow makes it easy to identify. These large trees have long, flexible branches and twigs that are covered with thin leaves that point downward. With its vertical branches and twigs, it can look as if the tree has curtains of leaves that extend all the way to the ground.

HOW TO SPOT

Height: 30 to 50 feet (9 to 15 m)

Leaves: Narrow, blade-like leaves; 4 inches (10 cm) long

Seeds: Tassel-like catkins; 1 to 2 inches (2.5 to 5 cm) long

Range: Western Asia, Europe, and North America

Habitat: Riverbanks and stream banks

AMERICAN HORNBEAM *(CARPINUS CAROLINIANA)*

The American hornbeam tree gets its name from its strong trunk and tough, heavy wood. The blue-gray bark is smooth and can have ripples that look like muscles. The hornbeam's papery winged samaras hang in clusters that look a little like pine cones. The tree's leaves are bright green throughout the summer and turn yellow, orange, or red in the fall.

FUN FACT

Other names for the American hornbeam include blue beech and water beech.

HOW TO SPOT

Height: 20 to 40 feet (6 to 12 m)

Leaves: Pointed ovals with toothed edges; 3.5 inches (9 cm) long

Fruit: Clusters of winged fruit; 1.2 inches (3 cm) long

Range: Eastern North America

Habitat: Forests, riverbanks, and swamps

COMMON WITCH HAZEL
(HAMAMELIS VIRGINIANA)

The common witch hazel is a small, shrub-like tree that often has a leaning trunk and crooked branches. Its broad, oval-shaped leaves have wavy edges and turn bright yellow in the fall. The common witch hazel produces seed capsules that look like tiny, brown flowers. When the capsules ripen, they split open and eject the seed.

HOW TO SPOT

Height: 10 to 25 feet (3 to 7.5 m)

Leaves: Broad ovals with wavy, toothed edges; 4 inches (10 cm)

Fruit: Small, woody, blossom-shaped capsules

Range: Eastern North America

Habitat: Riverbanks, stream banks, and rocky slopes

GLOSSARY

canopy
Leaves and branches that spread from the top of a tree.

catkin
A cluster of tiny flowers shaped like a tassel or cat's tail.

climate change
A significant and long-lasting change in the Earth's climate and weather patterns.

coniferous
Trees that have needles or scale-like leaves and produce cones.

crown
The top of a tree which is made up of branches.

cultivar
A specific variety of plant developed by people.

deciduous
Trees that drop their leaves at least once a year, often in fall or winter.

frond
A large leaf with many divisions.

leaflet
A smaller leaf that's a part of a larger compound leaf.

lobe
A rounded or wavy segment of a leaf.

resin
A sticky substance secreted through the bark of coniferous trees.

samara
A winged fruit or seed.

sap
The fluid part of a tree that circulates through the tree. Produced by both deciduous and coniferous trees.

TO LEARN MORE

FURTHER READINGS

Daniels, Patricia. *Trees.* National Geographic Kids, 2017.

Jose, Sarah. *Trees, Leaves, Flowers & Seeds: A Visual Encyclopedia of the Plant Kingdom.* DK Publishing, 2019.

Socha, Piotr. *Trees: A Rooted History.* Abrams, 2019.

ONLINE RESOURCES

To learn more about trees, please visit **abdobooklinks.com** or scan this QR code. These links are routinely monitored and updated to provide the most current information available.

PHOTO CREDITS

ABDOBOOKS.COM

Published by Abdo Publishing, a division of ABDO, PO Box 398166, Minneapolis, Minnesota 55439. Copyright © 2021 by Abdo Consulting Group, Inc. International copyrights reserved in all countries. No part of this book may be reproduced in any form without written permission from the publisher. Abdo Reference™ is a trademark and logo of Abdo Publishing.

Printed in the United States of America, North Mankato, Minnesota.
082020
012021

 THIS BOOK CONTAINS RECYCLED MATERIALS

Editor: Alyssa Krekelberg
Series Designer: Colleen McLaren
Content Consultant: Dr. David Neale; Department of Plant Sciences; University of California, Davis

Library of Congress Control Number: 2019954357
Publisher's Cataloging-in-Publication Data
Names: Debbink, Andrea, author.
Title: Trees / by Andrea Debbink
Description: Minneapolis, Minnesota : Abdo Publishing, 2021 | Series: Field guides for kids | Includes online resources and index.
Identifiers: ISBN 9781532193071 (lib. bdg.) | ISBN 9781098210977 (ebook)
Subjects: LCSH: Trees--Juvenile literature. | Trees, Multipurpose--Juvenile literature. | Field guides--Juvenile literature. | Reference materials--Juvenile literature.
Classification: DDC 582--dc23